PLATES

RESTLESS EARTH

EARTHW●RKS

PLATES

RESTLESS EARTH

ROY A. GALLANT

BENCHMARK BOOKS

MARSHALL CAVENDISH
NEW YORK

PLATES

RESTLESS EARTH

For my remarkable Russian friend, Alex Rudoy

Series Consultants:

LIFE SCIENCES AND ECOLOGY

Dr. Edward J. Kormondy
Chancellor and (professor emeritus) of Biology
University of Hawaii—Hilo/West Oahu

PHYSICAL SCIENCES

Christopher J. Schuberth
Professor of Geology and Science Education
Armstrong Atlantic State University
Savannah, Georgia

Benchmark Books
Marshall Cavendish
99 White Plains Road
Tarrytown, NY 10591-9001

Library of Congress Cataloging-in-Publication Data
Gallant, Roy A.
  Plates : restless earth / by Roy A. Gallant.
     p. cm. — (Earthworks)
Summary: Discusses plate tectonics, the theory that the surface of Earth is always
moving, and the connection of this phenomenon to earthquakes and volcanoes.
Includes bibliographical references and index.
  ISBN 0-7614-1370-7
  1. Plate tectonics--Juvenile literature. 2. Plate
tectonics—Environmental aspects—Juvenile literature. [1. Plate
tectonics. 2. Earthquakes. 3. Volcanoes.] I. Title. II. Series.
  QE511.4 .G355 2003
  551.1'36—dc21   2002000915

Photo research by Linda Sykes Picture Research, Hilton Head, SC

Cover: Bardarson/Earth Scenes
The photographs in this book are used by permission and through the courtesy of:
Breck Kent/Earth Scenes: 2–3, back cover; Krafft/Hoa-qui/Photo Researchers: 8; Corbis:
15; Bettmann/Corbis: 19; Natural History Museum, London: 27, 28; United States
Geological Survey: 30; Karen L. Von Damm, Ph.D., Professor of Geochemistry, and Earth,
Oceans and Space Complex Systems Research Center/EOS/Earth Sciences University of
New Hampshire: 32; Photo by Yvonne Malcolm-Coe, supplied by Celia Nyamweru: 40,
back cover; Space Images: 50; George Holton/Photo Researchers: 54 (top); Frans Lanting/
Photo Researchers: 54 (bottom), back cover; Paddy Ryan/Earth Scenes: 55, back cover;
Roy Gallant: 56–57, back cover, 59; Davd Halpern/Photo Researchers: 58; AFP/Corbis: 65.

Series design by Edward Miller.

CONTENTS

INTRODUCTION

The people of Iceland are constantly reminded that Earth is not such a "solid" planet after all. Visit the Krafla volcano in northeastern Iceland and you will find ever-widening cracks in the ground, with new ones breaking open all the time. As the cracks appear, the ground sometimes rises 3 to 7 feet (1 to 2 meters) and then suddenly drops back down again. Such rifting, as it is called, nearly always signals that Krafla is about to erupt. Volcanoes the world over are fiery proof that Earth's surface features are ever changing.

JOURNEYING INTO THE PAST

If you had a time machine and could replay a speeded-up version of Earth's history, you would see that the continents were once in places very different from their locations today. You could watch Saudi Arabia being torn from northeastern Africa and a gigantic ditch forming in between. Later you would see water flooding into the ditch to create what today is the Red Sea. Then, by using your time machine to peer into the future, you could watch southern California split off of the North American continent and move into the Pacific Ocean along that great crack in Earth's crust called the San Andreas Fault. You could also see the country of Indonesia sinking ever so slowly before it disappears beneath the waves of the South Pacific Ocean.

The idea that the continents drift about like great rafts of stone floating in a sea of molten rock astonished geologists of the early 1900s. Most denied that such a thing could happen. But over the years, researchers began to find clues that, indeed, it could, and they proposed a seemingly wild idea called continental drift. By the 1960s, research ships had collected surprising evidence

Red hot lava and clouds of ash explode out of Iceland's Heimaey Island. Iceland sits atop the Mid-Atlantic Ridge, a long crack in Earth's crust that runs down the middle of the Atlantic Ocean and spills out magma from deep within the planet.

that the seafloor of the Atlantic Ocean was widening and pushing Europe and North America slowly apart. Scientists also had discovered why the continents were on the move. They were riding atop enormous flat plates of stone. A new branch of earth science called *plate tectonics* arose from these discoveries. The term *tectonics* means "to build." It has unified the study of Earth by drawing together sciences as varied as paleontology, the study of fossils; seismology, the study of earthquake waves; and geophysics, the study of Earth's interior. It has given us answers to questions that scientists had wondered about for centuries—such as why earthquakes and volcanic eruptions occur most often in certain areas of the world, and how great mountain ranges like the Alps and Himalayas were formed.

Heat and pressure in Earth's deep interior are the driving forces that push the plates around and alter the planet's geography from one geologic age to another. But just how much heat does Earth have and how rapidly is it being lost? These are just two of the many questions geophysicists would love to be able to answer. Due to our scant knowledge of Earth's interior, the solutions are often elusive.

Unfortunately, we don't have a time machine, so we continue to poke and probe Earth's complex interior with the tools we have. Earthquake waves, which pulse and shiver through the planet, have proved valuable messengers. Scientists monitor and interpret their activity because they help reveal many of the mysteries that lie in Earth's churning depths.

ONE

EARTH'S SHAKY ROOTS

Earth is a restless planet. Great slabs of rock called plates make up the planet's crust, or *lithosphere*, and support the continents and sections of the ocean floor. Forces deep within Earth keep the plates moving about, crashing into each other, buckling up mountains, or elsewhere forcing huge sections of crustal rock down hundreds of miles into the planet's hot interior. All of this activity produces two of our most feared calamities—earthquakes and volcanoes. Long shrouded in mystery, both have caused devastation throughout human history. But they extend much further back in time. They have shaped and shaken the planet to its roots for about 4.6 billion years.

Although it took you only a minute to read those sentences and learn that Earth teeters on enormous plates of stone, it took the world's greatest thinkers more than two thousand years to come to that understanding.

It wasn't until after countless false starts, struggles through the bogs of superstition, and futile attempts to read the Bible as a scientific text that scientists of the past fifty years came to build a solid scientific basis for the theory of plate tectonics.

WHAT HELD UP EARTH?

The yearning to unravel the mystery of Earth's shudders and rumblings led ancient scholars to pose even larger questions, such as what held Earth in place among the stars? Around 500 B.C. people commonly believed that Earth was motionless in space and that everything else in the heavens circled around it. So they reasoned that some kind of support for Earth must be necessary. But what? They also thought that the occasional shaking of Earth must be caused by the motion of that system of support. It was around this time that the Greek philosopher Anaxagoras taught that Earth was flat and floated on a cushion of air. Earthquakes, he believed, were caused when air above the planet rushed down and plunged against the air cushion beneath.

The great philosopher Aristotle, who was born in 384 B.C., taught that earthquakes were caused by winds blowing into caverns within the planet. When the pressure of the trapped air became strong enough, the air escaped forcefully and caused Earth to tremble.

Earlier, another philosopher named Democritus had supposed that hollows within Earth collected water, and when the water sloshed back and forth, the planet shook. Quite possibly the many sea caves found throughout the Greek islands suggested these notions of a hollow Earth.

BUFFALOES, FROGS, AND SERPENTS

Long after the time of the ancient Greek thinkers, people in different parts of the world still felt that Earth must have a support of some kind. Lacking facts, people turned to their imaginations. The folklore of many different cultures around the world relates that different kinds of animals provided Earth with its shaky support.

At various times, certain people ranging from the Pacific Islanders to Eastern Europeans have supposed that a gigantic buffalo carried Earth on its back. Whenever the beast shifted its weight from one foot to another, there was an earthquake. The Algonquian Indians thought the animal was a giant tortoise. In Mongolia it was a frog; in the Celebes, a hog; in Persia, a crab. In India seven serpents took turns supporting the planet. Each time a new serpent took over, Earth trembled. The geologist L. Don Leet relates this tale in his 1948 book *Causes of Catastrophe*:

> According to the Masawahilis of East Africa, a monster fish called the Chewa swims in the world ocean and carries on its back a stone upon which there is a cow which carries Earth on one of its horns. Whenever the cow shifts Earth from one horn to the other, there is an earthquake.

More recently, earthquakes have also been blamed on the changing positions of the planets. In astronomy, we refer to the "alignments" of the planets, when most of the planets are stretched out on the same side of the Sun. This was thought to produce an especially strong gravitational tug on Earth that brings on earthquakes. The English astronomer John Gribbin popularized that notion in one of his books published in the 1990s. He later apologized and retracted his statement. From time to time there are planetary alignments, of course, but they are not the cause of earthquakes. The notion is every bit as false as Aristotle's vast caverns puffed up with air.

SACRIFICES AND THE GREAT GODDESS PELE

The grinding and crushing of Earth's plates against one another not only make the planet tremble, but also trigger events beneath the crust that send great volumes of molten rock surging up as volcanoes. The hot rock, called *magma*, then gushes out of a volcano's vent as *lava*. But that's not how people of old explained volcanic eruptions.

The Romans in ancient times supposed that the fiery outbursts were the work of their god of fire and metalworking, Vulcan. A volcano erupted whenever Vulcan fired up his forges to make weapons for Mars, the god of war. So these mountains were called "vulcanoes," later changed to our word volcanoes.

The Greek geographer Strabo, born around 64 B.C., agreed with Aristotle's winds theory about earthquakes. But Strabo took Aristotle's thinking one step further by saying that when pressure within the wind-packed caverns became great enough, volcano "safety valves" were activated. Pressure was released, supposedly, so that Earth didn't blow itself to bits.

Both Strabo and Aristotle were searching for the natural causes of volcanoes. As with earthquakes, others, however, have turned to superstition. Cultures including the Aztecs, Mayas, and Incas believed that a god or goddess supposedly became angered by the behavior of the people and threatened them with destruction by volcano. This became a common explanation shared by many other cultures as well. To soothe the angered god, the high priests—the powerful politicians of old—offered human sacrifices by throwing children or young adults into the fiery crater to appease the god.

Exactly how throwing a child into a volcano was supposed to calm an angry god was never exactly explained, but the high priests nevertheless got their way. Although humans are no longer sacrificed, some cultures still do what they consider the next best thing. On the Pacific island of Java, tribal leaders to this day throw live chickens into the crater of Mount Bromo once a year to persuade the volcano not to erupt.

Also, to this day the goddess Pele of the Hawaiian Islands is alive and well, at least in the minds of her followers. Whenever lava from an erupting volcano threatened to engulf a Hawaiian village, a member of the royal family would be called on to calm Pele's anger and so stop the flow. In August 1881, the volcano Mauna Loa threatened the city of Hilo. Princess

This engraving made in the 1600s illustrates the belief that Earth's many volcanoes were all connected within the planet and that the molten rock continually flowed up to the surface. Icelandic people who lived nine hundred years ago thought of their famous volcano, Hekla, as the entrance to Hell. Whenever it erupted, people claimed to hear loud wailing, mournful cries of lost souls, fearful howling, weeping, and the gnashing of teeth.

Ruth Keelikolani rushed there and strode to the edge of the advancing molten rock. Chanting in her ancient language, she made offerings of silk scarves to Pele and poured brandy onto the advancing lava. The next morning the lava flow stopped, and the village was saved.

In 1955 an outpouring of lava from the volcano Kilauea threatened the Hawaiian village of Kapoho. People gathered at the advancing pool chanting and throwing offerings of food and tobacco into the lava. Once again, the flow stopped short, although it destroyed another village along with 6 square miles (16 square kilometers) of farmland.

TWO

THE CREEPING
OF THE CONTINENTS

Surely no one would believe this new upstart, Alfred Wegener, in their midst. The world-famous geologists attending this important meeting would laugh at him when he tried to convince them of his wild idea—that throughout Earth's history the solid rock continents with their deep granite roots have actually drifted about like massive ships of stone and are still moving to this day. To see for themselves, Wegener argued, all they had to do was observe how South America and Africa form an almost perfect fit when placed side by side like two pieces of a jigsaw puzzle. That couldn't be a coincidence. The shorelines of North America and Europe also matched fairly well, although less obviously than South America and Africa. How could that be if the continents were not once joined, Wegener asked? They scoffed when he claimed that the solid ocean floors also take part in this movement by stretching wider.

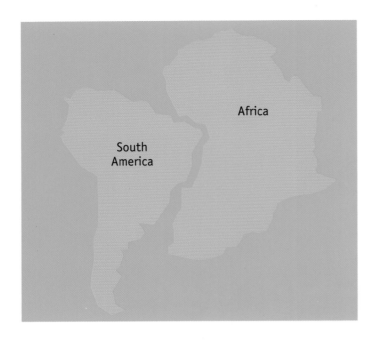

South America and Africa form a neat fit, which suggested to some that the continents were once joined but later drifted apart. The idea finally gained favor when it was shown that certain rock formations span the present boundaries of the two continents.

Further, he told them: "Gentlemen, it is just as if we were to refit the torn pieces of a newspaper [by matching their edges] and then check whether the lines of print run smoothly across. If they do, there is nothing left but to conclude that the pieces were in fact joined in this way."

Wegener was anxious as he addressed the January 1912 meeting of the German Geological Association. After all, he was only a meteorologist addressing important geologists. Unknown to him at the time, he was prying open the door of a new science, plate tectonics, which would grow out of his theory of *continental drift*.

AN ISLAND THAT DRIFTED AWAY

Around 1910 Wegener was working as a meteorologist with a Danish scientific expedition in Greenland. Eighty-seven years earlier the British geographer Sir Edward Sabine had discovered and plotted a small island, later named Sabine Island, off the east coast of Greenland. In 1869 another group of geographers had replotted the island's position and found that it was a quarter of a mile (nearly half a kilometer) west of where Sabine had originally placed it on his maps. To Wegener's thinking, this error was too great to be accounted for by faulty measurement.

Could it be, he wondered, that the island, rather than the geographers, was "wrong"? In other words, could Sabine Island have moved as much as a quarter of a mile between 1823 and 1869? Wegener took new measurements and found that since 1869 the island had continued its westward shift by a little more than an additional half-mile (nearly a kilometer).

Later Wegener checked the positions of other arctic landmasses and concluded that all of them were drifting westward at different speeds. America supposedly was moving away

In the early 1900s when the German scientist Alfred Wegener said that the continents drift about like gigantic stone rafts in a sea of molten rock, most geologists scorned the idea. Wegener did not live to see his theory of continental drift accepted.

from Europe at the rate of a quarter of an inch (almost half a centimeter) a day. Although his estimate proved to be too fast, his data and observations spawned the development of his drifting continents theory. So he set out to collect more evidence than just the matching coastlines. Even more convincing, he said, was that the same types of plant and animal fossils were found along the facing coasts of Africa and Brazil, of Europe and of North America, and of Madagascar and India.

Some geologists had suggested that land bridges once linked those now distant places and so enabled animals of long ago to travel back and forth between the now separated regions. Wegener did not accept the land bridge argument, and few scientists today think they ever existed. Instead of these

About 250 million years ago, an enormous supercontinent called Pangaea made up Earth's surface. Over time, Pangaea was broken apart by the movement of rigid "rafts" of solid rock afloat on a sea of puttylike rock. The present continents continue to move about, pushed this way and that by forces deep within the planet.

terrestrial links, Wegener envisioned an original supercontinent called Pangaea, meaning "all Earth," surrounded by a world ocean called Panthalassa, meaning "all sea." Then, about 200 million years ago, Pangaea began to break apart into a northern half called Laurasia and a southern half called Gondwana. Further fractures in the landmasses and their subsequent drifting moved the continents to their present positions, Wegener said.

WEGENER'S "IMPOSSIBLY ABSURD" THEORY

His idea was so revolutionary that Wegener anticipated vigorous criticism. But it is doubtful that he braced himself for just how severe it would be. During a 1926 meeting of the American Association of Petroleum Geologists, the assembled experts tore into Wegener's theory, determined to shred every scrap of evidence he had produced. They especially attacked Wegener's weaker arguments. For example, he had incorrectly claimed that the westward drift of the Americas was caused by gravitational forces of the Sun and Moon, a notion that he could not support. Then they launched a personal attack by saying that he had no business introducing theories in a scientific area where he was not an expert. He was a mere meteorologist, they sneered.

As unpleasant as all this criticism was, Wegener did not let it defeat him. Determined as ever to prove his theory, he continued to research and write. He also went on two more expeditions to Greenland. While on an expedition there in 1930, Wegener died, apparently freezing to death during a mission to deliver supplies to a research camp far inland after winter had set in. It would be more than thirty years before Wegener's continental drift theory got the recognition it deserved.

Although Wegener was the first to assemble convincing evidence for continental drift, he was not the first to propose the notion that the continents move. As early as 1596, the Dutch mapmaker Abraham Ortelius suggested that the Americas were "torn away from Europe and Africa . . . by earthquakes and floods." In 1658 the French monk, R. P. François Placet, declared that the neat fit of South America and Africa strongly suggested that the Old and New Worlds broke apart after Noah's Flood. In 1756 the German theologian Theodor Lilienthal echoed the same view. Forty-four years later the German explorer Alexander von Humboldt said that the Atlantic Ocean basin was a huge river valley somehow associated with the biblical Flood. Then in 1858 the Italian geographer Antonio Snider-Pellegrini also advanced the idea that all of the continents were once joined and later broke apart.

By the late 1800s Sir George Darwin, son of the naturalist Charles Darwin, had proposed that the Moon was torn out of Earth's side by the gravitational attraction of a passing star. As Earth adjusted to its loss of mass, its land forms were redistributed, resulting in the breakup of continents, he explained.

Darwin's idea of the Moon's birth enjoyed popularity for fifty years or so, until astronomers of the 1900s favored the theory that the Moon was formed along with Earth out of the great wheeling disk of matter cast off by the Sun some 4.6 billion years ago. But then in the late 1900s astronomers changed their minds again by proposing that the impact of a giant asteroid

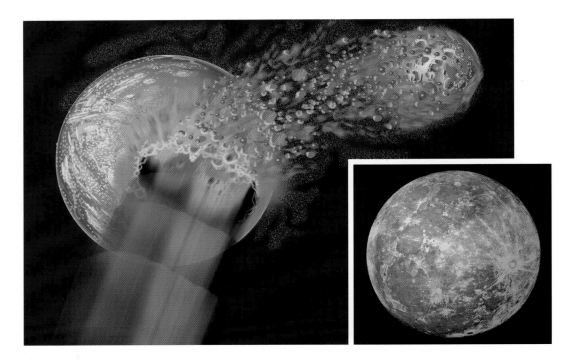

Although details of the theory are yet to be worked out, most astronomers now think that the Moon was splashed off Earth when the planet was struck by a huge planetesimal (a bit smaller than Mars) 4.5 billion years ago. Within some ten hours, the smashed-out matter formed a sphere that became our Moon.

smashing into the planet caused a great splash of molten matter which eventually solidified to form the Moon. Such a cataclysmic event certainly would have rearranged whatever continental blocks might have existed at the time.

About the time Wegener was working out his theory of continental drift, two Americans, Frank B. Taylor and Howard B. Baker, were forming their own ideas on the subject. Taylor cited as evidence mountain chains, including those along the west coasts of North and South America, which would continue unbroken if the continents were pushed back together again. Baker pointed out that such a squeeze would also line up the Appalachian Mountains on the east coast of North America with the old

Caledonian mountain chain that runs through Scandinavia, Scotland, and Northern Ireland. Wegener also had cited the wrinkling up of mountains as evidence for continents ramming each other, the collisions buckling up the mighty Alps and Himalayas. But despite how hard they tried, Wegener, Taylor, and Baker could not convince the leading geologists of the day, such as England's highly respected geophysicist Sir Harold Jeffreys, who dismissed the notion outright.

It wouldn't be until the 1950s and 1960s that studies of the Atlantic Ocean's floor were to prove Wegener right. The evidence was Earth's

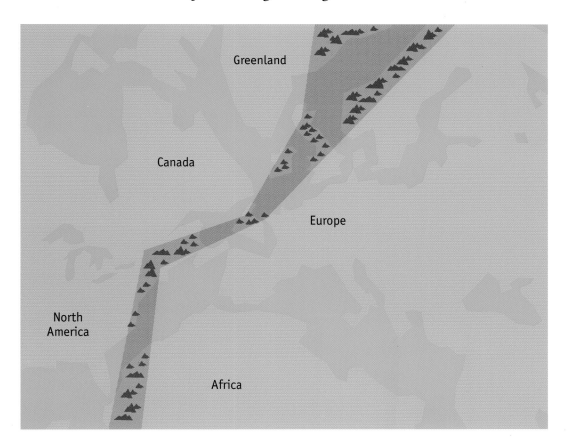

If the continents were fitted together, the Appalachian and Caledonian mountains would form a continuous chain. The existence of such an ancient chain suggested to geologist Howard B. Baker that the continents were, indeed, once joined.

Plates

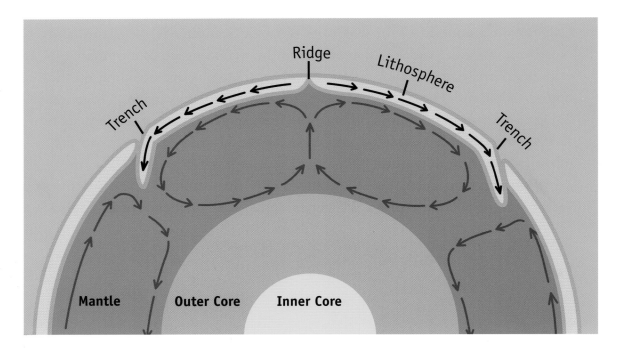

Ridge

Trench

Lithosphere

Trench

Mantle **Outer Core** **Inner Core**

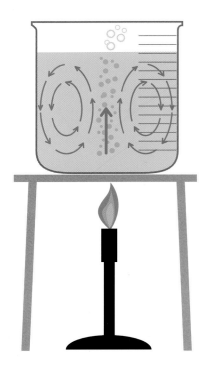

The geologist Arthur Holmes looked to convection cells of hot rock beneath Earth's crust circulating and so causing the solid crustal rock to move about. Like water heated over a flame, molten rock heated by the extremely hot core region became lightweight and rose toward the surface. As it did, it cooled, became heavier, and sank back toward the core. Holmes felt that the pressure of the circulating cells of hot rock would be strong enough to split continental land masses apart.

changing magnetic field over time, a record of which was preserved in the ancient rocks of the ocean floor. This evidence helped establish that continents drift, altering the magnetic field along the way. It also raised the even more basic notion that the planet's crust, or lithosphere, is made up of about six giant rock platforms and about a dozen smaller ones, all kept in slow motion by the great exchange of heat beneath the lithosphere and in the rock still deeper down.

Wegener had finally been vindicated. All the time, his disbelieving colleagues had unknowingly been floating on gigantic plates of stone whose existence they refused to acknowledge.

THREE

SEAFLOOR SPREADING

Around 1929, the English geologist Arthur Holmes enlarged on one of Wegener's ideas. He proposed that hot fluid rock beneath Earth's lithosphere circulated in mammoth convection currents. As hotter and, therefore, lighter-weight rock rose toward the crust it cooled and grew heavier before sinking back down again. Holmes imagined great cells of molten rock circulating like conveyer belts in Earth's mantle, that huge layer stretching for nearly 2,000 miles (3,220 kilometers) down to Earth's core. He said that the pressure of the rising molten rock would be great enough to split apart the crust and then carry the pieces off in opposite directions. But Holmes's idea received little attention at the time.

It wasn't until the 1950s that geologists and geophysicists began to reconsider Wegener's and Holme's continental drift theories. That renewed interest was sparked by studies not of the land but of the deep ocean floor made possible by new scientific devices, including echo sounders and magnetometers.

HOW MOUNTAINS MAKE NEW SEAFLOOR

As one geophysicist has remarked, "Until a century ago about the only thing we knew about the floors of the oceans was that they were very deep." But that changed dramatically over the last century when various research vessels discovered huge undersea ridges snaking across the floors of the world's oceans. The first such ridge was discovered in the 1870s by the British research ship HMS *Challenger*. The ridge runs for 10,000 miles (16,000 kilometers) down the middle of the Atlantic Ocean and pokes up above the waves as the volcanic land masses of Iceland in the north and Tristan da Cunha off the tip of Africa in the south. Called the Mid-Atlantic Ridge, it did not exist 165 million years ago, but then again neither did the Atlantic Ocean.

HMS Challenger's 1872 to 1876 voyage marked the first expedition to reveal the mysteries of the deep seafloor. The voyage took oceanographers to every ocean except the Arctic.

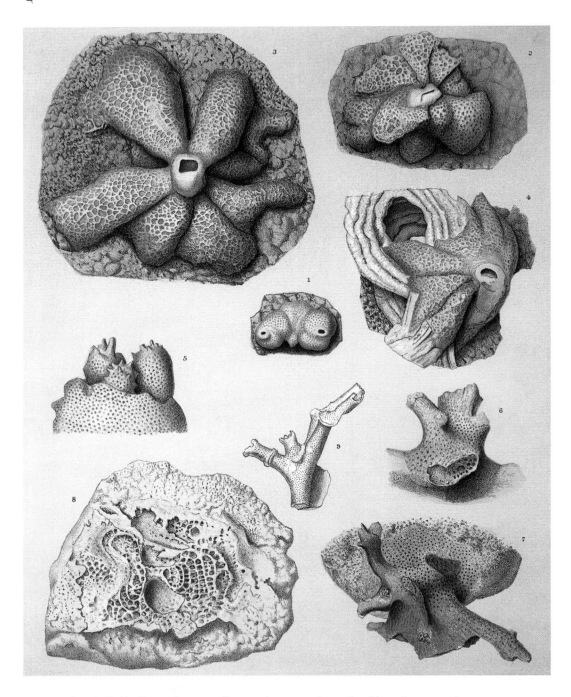

Among the HMS *Challenger's* many discoveries were deep-dwelling tiny organisms called foraminifera.

MEASURING THE OCEAN DEPTHS WITH ECHOES

An echo sounder works by sending sound wave "pings" from a research ship to the ocean floor. When the sound waves strike the bottom, they bounce back to the ship as an echo. While a delicate receiver records the echo, a clock measures the interval between each ping transmission and its return as an echo. Because the speed of sound through seawater is accurately known—4,860 feet (1,480 meters) per second—the ocean depth can be measured at any point. Echoes can also be used to sweep over an area of the seafloor and so provide a general map of its features. Another method, seismic surveying, tells a researcher about the kinds of rock making up the ocean bottom. Sound waves from a charge exploded underwater penetrate the 2,000 feet (610 meters) of sediments and travel through the rock floor to a second ship that receives the signal. Since scientists know the speeds of sound through different kinds of rock, measuring the travel times of sound waves sent through the ocean floor reveals the different types of rock found there.

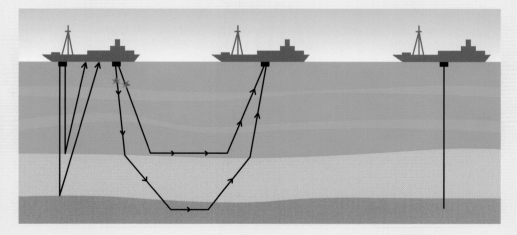

Three ways of studying the seafloor: Echo sounding (*left*) measures the time it takes for a sound pulse to bounce off the seafloor and return to the ship. Seismic surveying (*center*) reads the kinds of rock material composing the seafloor. Core sampling (*right*) fires a long hollow tube into the seafloor sediments and then draws the tube up so its contents can be studied in the laboratory.

By 1956 American oceanographers Maurice Ewing and Bruce C. Heezen had shown that the various ocean ridges were connected as a vast underwater mountain range extending around the globe over some 41,000 miles (66,000 kilometers), and reaching an average height of about 14,760 feet (4,500 meters).

Then in the early 1960s Harry H. Hess of Princeton University and Robert S. Dietz of the U.S. Coast and Geodetic Survey revived and expanded Holmes's

The research vessel *Glomar Challenger* provided final proof of seafloor spreading. In 1968 it criss-crossed the Mid-Atlantic Ridge between South America and Africa, drilling core samples at many locations along the way. The differing ages of the samples—older near the edges of the continents, younger near the ridge's rift valley—showed that seafloor spreading was not theory but fact.

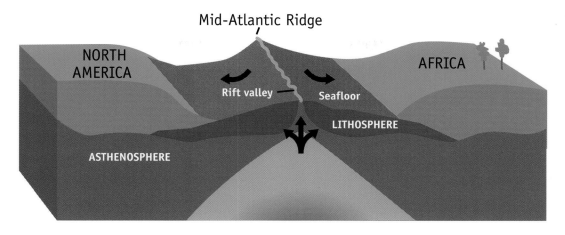

The ocean floor spreads as new floor material as magma, or molten rock, wells from below and spills out of the Mid-Atlantic Ridge. Before the ridge began to spread the seafloor apart, all the land masses now separated by the Atlantic Ocean were still joined. The great ridge did not exist 165 million years ago.

thermal convection theory and established the idea of seafloor spreading. While Holmes's theory was hardly taken seriously, new evidence made Hess's revision of the theory much more appealing. Mid-ocean ridges, deep sea trenches, arcs of islands such as those forming Japan and the Philippines, distinctive magnetic patterns in the ocean floor rocks, and fault patterns all indicated that heat currents deep within the planet were continuously at work changing features at the surface.

Oceanographers once thought that the clay, mud, and other materials that washed off the continents should be spread rather evenly on the ocean floor. However, moving east or west from the Mid-Atlantic Ridge, they found that the sediments are thickest near the continents and only thinly dispersed near the mid-ocean ridge. They wondered why. The answer came with the discovery of a rift valley along the top of the mountain chain. From time to time molten rock from deep within the mantle wells up through the rift valley. As it spills out, the newly forming lava forces the valley walls

Called "black smokers," hydrothermal vents boil up out of the seafloor in many parts of the oceans. This one is at a depth of 9,300 feet (2,834 meters) along the East Pacific Rise. The fountain of ejected fluids is mainly made up of iron and sulfur compounds at an unusually high temperature of 760 degrees Fahrenheit (405 °C).

apart and so widens the valley floor. The molten rock then cools and forms new seafloor rock, called *basalt*, along both edges of the ridge. That explains why the sediments are shallower closer to the ridge and thicker farther away. The newer seafloor near the ridge has had less time to collect sediments.

MAGNETISM AND FOSSILS

A magnetometer is an instrument that can read the magnetism of rocks. In addition to the echo sounder, it was another important device that helped provide evidence of seafloor spreading. Many rock types become magnetized when they are formed. For instance, sedimentary rocks such as sandstone and shale begin to be magnetized when they are still soft sediments, before they solidify. Lava acquires its magnetic charge as it cools and hardens. In both cases, tiny mineral crystals in the rock line up with the planet's magnetic field and so become internal compasses frozen in time. Reading those tiny compasses reveals the directions of the planet's North and South Poles at the time the sediments were deposited or the lava hardened into rock.

Hundreds of magnetometer readings of the Atlantic's rock floor showed a series of stripes, like those of a zebra, running parallel to both sides of the mid-ocean ridge. Researchers reasoned that if, indeed, new ocean floor was formed by the outpouring of magma along the mid-ocean rift valley, then the magnetic stripes on one side of the rift should be a mirror-image of the stripes on the opposite side. They were. Surprisingly, the various patterns of magnetism frozen in the seafloor rocks showed that the north and south magnetic poles have flip-flopped about 170 times over the past 76 million years or so—the northern pole switching places with the southern pole. Further proof of seafloor spreading came from measuring the age of the stripes. If the seafloor was actually spreading, then the stripes of rock nearest the rift should be youngest because they were laid down by the most recent outpourings of lava. Likewise, those progressively farther from the rift

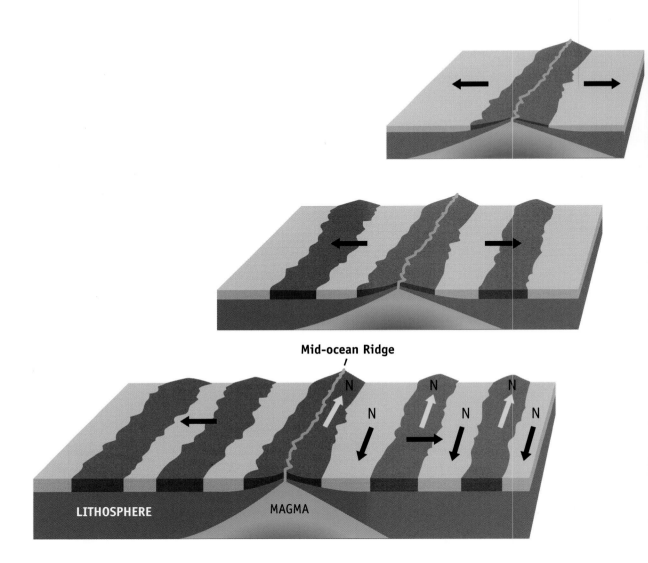

Mid-ocean Ridge

LITHOSPHERE   MAGMA

Like zebra stripes on the seafloor, these magnetic stripes mark seafloor spreading at various times in Earth's past. The stripes show that the planet's magnetic poles have reversed time and time again. Magnetic mineral grains contained in lava outpourings from the Mid-Atlantic Ridge align in a north-south direction and serve as frozen compass needles when the lava cools and solidifies. The colored stripes indicate a north polar alignment while the dark arrows indicate periods when the planet's magnetic poles did a flip-flop.

should be older. Again, they were. Two English scientists, Drummond Matthews and Frederick J. Vines, played major roles in explaining the meaning of the stripes associated with seafloor spreading.

In further support of Wegener's findings, the fossils of certain animal types were found on continents that were widely separated. For instance, the fossil bones of a 240-million-year-old reptile about the size of a dog, called *Lystrosaurus*, were found in Antarctica, Africa, India, China, Russia, and Mongolia. Since Lystrosaurus was a land dweller, the only way the animal could have inhabited those now distantly separated regions was if they were once joined as a single landmass.

The new science of plate tectonics was born when it was learned that the undersea ridges form the edges of the massive plates. By the mid-1960s virtually every geologist had come to accept continental drift as fact. The once unreachable ocean floor was slowly giving up its secrets, providing a more complete understanding of the planet as a whole.

FOUR

HOW
PLATES
MOVE

In 1968 the Canadian geophysicist J. Tuzo Wilson asserted that "the earth, instead of appearing as an inert statue, is a living, mobile thing." That statement contradicted the old, tightly held notion that the planet was a solid, motionless body. Today we know that Earth's outer crustal rock, the lithosphere, is a mosaic of solid plates that slide over the hot, mushy asthenosphere, which surrounds the lower mantle. Wherever the plates collide along their edges, mountains may be thrust up, earthquakes might send pulses through the planet, and volcanoes could blow their tops.

THREE WAYS PLATES MOVE

The ways in which plates collide determine what occurs at and near their edges. One thing we can be sure of is that something always happens. All the plates move slowly, on the average about as fast as your fingernails grow. While some move less than 0.5 inch (1 centimeter) a year, others "speed" along at about 3.5 inches (9 centimeters) a year. Whether fast or slow, over hundreds of thousands of years plate movement continually shapes and reshapes the planet's land-forms and ocean basins. In some places, such as the Atlantic Ocean, it stretches and broadens the seafloor. In other regions, such as the Pacific Ocean, the seafloor is compressed by the westward movement of the North American and South American poles. Over still longer periods measuring several hundred million years, plate movement breaks up continents or smashes them together. All this activity occurs within the top 125 miles (200 kilometers) or so of Earth's upper layers.

DIVERGENT BOUNDARIES

One way plates move along their edges is by pulling away from each other. The lines along which separating plate edges meet are called *divergent boundaries*, one example being the Mid-Atlantic Ridge, where the seafloor expands between the North American Plate and the Eurasian Plate. The divergent boundary there runs right under Iceland and is splitting that small island in two. The rate of spreading along the Mid-Atlantic Ridge is about 1 inch (2.5 centimeters) a year, or 16 miles (25 kilometers) in a million years.

Another divergent boundary is the one between the South American Plate, the African Plate, and the Arabian Plate. The separation of these plates results in the widening of the Red Sea and the splitting away of East Africa along a huge crack in the crust called the Great Rift Valley. This widening along the rift valley may cause East Africa to be the site of Earth's next major ocean.

NORTH
AMERICAN
PLATE

EURASIAN
PLATE

Mid-Atlantic Ridge

CARIBBEAN
PLATE

ARABIAN
PLATE

COCOS
PLATE

AFRICAN
PLATE

East Pacific Rise

NAZCA
PLATE

SOUTH
AMERICAN
PLATE

SCOTIA PLATE

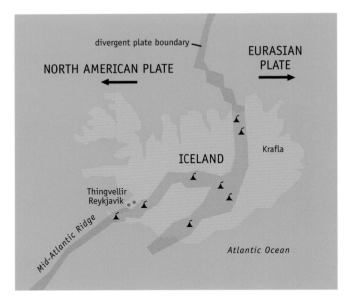

North American Plate ← divergent plate boundary → Eurasian Plate

ICELAND

Krafla

Thingvellir
Reykjavik

Mid-Atlantic Ridge

Atlantic Ocean

Left: Sitting atop that huge crack in Earth's crust called the Mid-Atlantic Ridge, Iceland is being split apart by seafloor spreading. This is causing the North American and Eurasian plates to split apart along a divergent boundary.

Below: An ancient volcanic rise, the Oldoinyo Lengai, in the East African Rift zone, may one day be covered by a new ocean. Water from the Indian Ocean may flood the region as the Arabian Plate pulls apart from the African Plate.

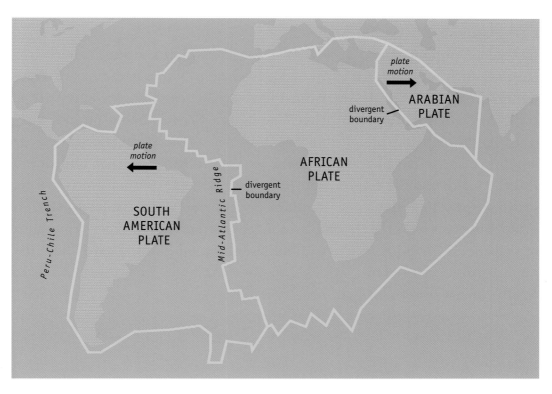

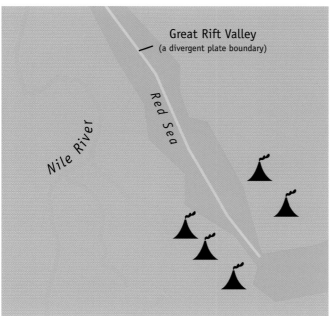

Great Rift Valley
(a divergent plate boundary)

Above: Divergent boundaries at the edge of the South American and African Plates and at the edge of the African and Arabian Plates are pulling the plates apart.

Left: Just as its big brother the Mid-Atlantic Ridge is spreading apart the seafloor of the Atlantic Ocean, the Great Rift Valley is spreading apart along the sea-floor of the Red Sea. And just as the once dry Atlantic Ocean basin was flooded beginning some 200 million years ago, massive flooding may occur in the Red Sea region, which would turn northeast Africa into an island.

CONVERGENT BOUNDARIES

Instead of pulling apart, a *convergent boundary* marks a zone where two plates crunch together. An example fairly close to home is the Nazca Plate ramming into the South American Plate. In this case, an oceanic plate is colliding with a continental plate. The Nazca Plate is being pushed eastward by outpourings of volcanic rock and the spreading of the seafloor at another major rift valley called the East Pacific Rise, another mid-ocean ridge. Meanwhile the South American Plate is being pushed westward by seafloor spreading along the southern part of the Mid-Atlantic Ridge. So what happens where these two plates meet? At the surface there is a wealth of volcanoes, earthquake activity, and mountain building. That is how the mighty Andes Mountains, which run the length of the western edge of South America, were formed.

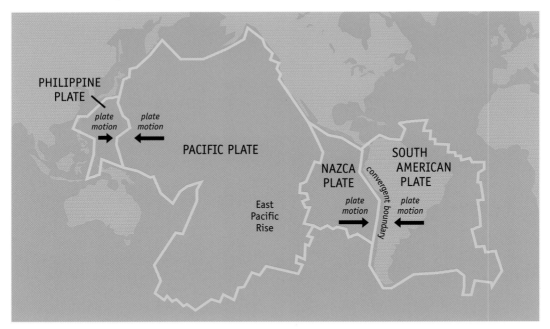

Two examples of convergent boundaries, where two plates ram one another: the boundary between the eastward-moving Nazca Plate and the westward-moving South American Plate, and the heavy, eastward-creeping oceanic Philippine Plate smashing into the westward-moving oceanic Pacific Plate. The eastern edge of the Pacific Plate has a feature called the East Pacific Rise, which is like the Mid-Atlantic Ridge with its rift valley in the Atlantic Ocean.

But something even more interesting happens deeper down along the common border of the colliding plates. The western edge of the lighterweight silicated rock of the continental South American Plate is riding up over the eastern edge of the heavyweight, basaltic rock of the oceanic Nazca Plate. This shoves the leading edge of the Nazca Plate down into the upper mantle, forming what geologists call a *subduction zone*. As the edge of the plate is forced down, it pulls a portion of the spreading basaltic ocean-floor crust with it. The result is the formation of a deep *oceanic trench*, known as the Peru-Chile Trench, along the western shore of South America.

As the edge of the Nazca Plate is bent and pushed down into the hot mantle, it is heated so much that it melts. This newly melted rock, now lightweight because it is hot, rises and melts its way up through the South American Plate. From time to time the fluid rock boils up to the surface and spills over the surrounding land as volcanic outpourings that continue to build up the Andes to this day. The Cascade Mountains of northern California, Oregon, and Washington are also formed in this way. Both mountain chains run parallel to trenches that lie next to the edges of the continents. It takes about 55 million

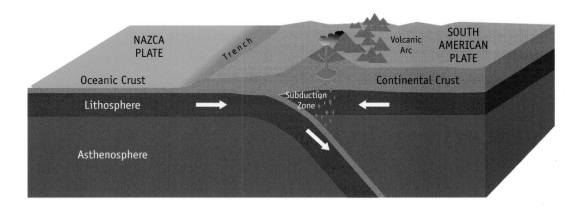

The west coast of the South American Plate is crunching against the east coast of the Nazca Plate. As a result, the heavier leading edge of the Nazca Plate is being forced down. Where the plates meet, at a point called a subduction zone, pressure melts the rock, which then rises and floods out as volcanoes.

years for newly formed ocean-floor crust to spread to the edge of one of its plates, become subducted and then melted. The life span of the continental crust is much longer, about 2.3 billion years.

Years ago, when the spreading of the seafloor became better understood, some geologists wondered if Earth could be expanding as new seafloor kept being formed. Since we are unable to detect any change in our planet's size by satellite observation, we can assume that as new seafloor crust is formed by magma upwellings from the mid-ocean ridge system, old seafloor is destroyed at about the same rate as it plunges down into the mantle along subduction zones.

TRANSFORM BOUNDARIES

Another way two plates can move is to rub and grind against each other along their common *transform boundary*. The area of contact may run from a few miles to hundreds of miles long, depending on the size of the plates involved. This kind of contact between plates produces a rupture line of crustal rock called a *fault*. California's famous San Andreas Fault is one such example and marks the border of the North American and Pacific Plates. Year after year, as the two plates push against each other along the 800-mile (1,300-kilometer) fault line, friction causes the rock to stretch. Eventually, the rock faces along the several-mile-wide contact zone reach their limit of stretching and they snap, like snapping your fingers, causing the ground to shudder along the fault line.

TRENCHES AND ISLAND ARCS

If we could drain all the water out of the Pacific Ocean, we would be able to see the enormous trenches formed by subduction that are carved into the ocean floor. A section of one, the Challenger Deep in the Mariana Trench, plunges more than 6 miles (10 kilometers) and is large enough to contain Mt. Everest, Earth's highest mountain.

In the case of the Nazca and South American Plates, a heavier oceanic plate rammed into a continental plate of relatively lightweight rock. But what happens when two heavyweight oceanic plates collide, such as the Philippine Plate and the Pacific Plate?

As you would expect, one plate is forced beneath the other, forming a subduction zone and a deep trench, and sections of the subducted plate melt. But instead of forming a continental mountain range, the hot rock melts its way up through the seafloor as a string of volcanoes running the length of the subduction zone. These undersea volcanoes continue to build in height until eventually they poke their tops above sea level and form a curved chain of islands called an island arc.

Examples are the Mariana Islands near the Mariana Trench, the Aleutian Islands off Alaska, and the islands of Japan. These island arcs are scenes of volcanic activity and strong earthquakes as the colliding plates interact. We can imagine the Pacific Ocean as a gigantic circular pond rimmed by a nearly continuous chain of volcanoes often called the Ring of Fire. Several segments of the ring are island arcs.

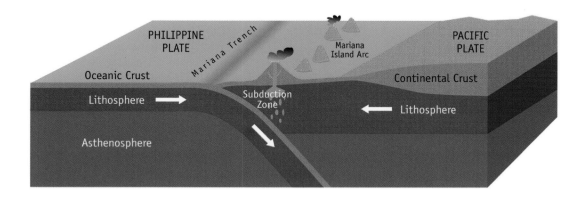

Where two oceanic plates collide, a subduction zone is formed, and where one plate plunges beneath the other, a deep trench results. Just beyond the trench upwelling molten rock forms an arc of islands, as in the Mariana Island Arc.

One of the most spectacular examples of plate tectonics is the Himalayan Mountains that tower between India and Tibet. The mountains were buckled up by the collision of two continental plates—the Indian Plate and the Eurasian Plate. Since both are made of lightweight continental rock, they crunched together and pushed up mountains. Most of the growth of the Himalayas and Tibetan Plateau has taken place in the last 10 million years and continues to this day. Presently the Tibetan Plateau has been shoved skyward to an average height of 15,000 feet (4,600 meters), greater than that of nearly any of the peaks of Europe's Alps.

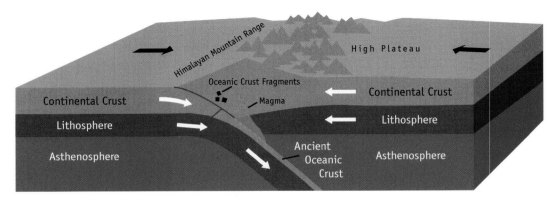

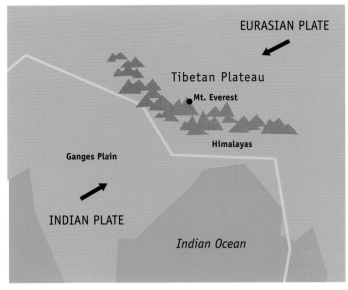

When two lightweight continental plates such as the Indian and Eurasian Plates collide, mountains are buckled up. This process of mountain building atop a subduction zone continues to this day (top), thrusting the Himalayan Mountains ever higher.

WHAT MOVES THE PLATES?

Even though we can detect the movement of plates with great precision—with the use of satellite measurements—we are less sure about just what makes them move the way they do. Until fairly recently, geologists felt that seafloor spreading is what sets the plates in motion by pushing some of them apart.

The old plate rock at the opposite edge of the plate is then subducted and pushed down into the mantle. The picture now seems to be reversed. It may turn out that the sinking subducted edge of a plate pulls the rest of the plate along with it in an action called "slab pull." It is not likely that the puzzle of plate movement will soon be solved because the forces involved—convection cells of magma within the mantle rock—are even more deeply buried than the plates themselves. Nevertheless, we can be certain that those forces, as mysterious as they now may seem, are ever at work changing the face of our planet.

FIVE

PLUMES AND HOTSPOTS

Since most of Earth's volcanoes are found along the edges of plates, how do geologists explain those volcanoes found near the middle of a plate? For example, the volcanic mountaintops of the Hawaiian Islands are more than 2,000 miles (3,200 kilometers) from the nearest plate boundary.

Hawaiian folklore provides one answer. More than a thousand years ago Polynesian mariners discovered and settled among the Hawaiian Islands. Over the following generations they developed a folklore about their home-land and its volcanoes. It reflects observations about the island chain that scientists have come to learn about only relatively recently: as the chain expands and new islands rise above the water, they always appear to the southeast. During their sea voyages, Hawaiians of long ago saw that there were differences in the amount of erosion and in soil and vegetation types from one island to the next. For example, Niihau and Kauai in the north-west appeared older than Maui and Hawaii in the southeast. These observations had been handed down from one generation to the next as the legends of Pele, the fiery goddess of volcanoes. They are so compelling they have even found their way into the U.S. Geological Survey's publication, *This Dynamic Earth*:

> Pele originally lived on Kauai. When her older sister Namakaokahai, the Goddess of the Sea, attacked her, Pele fled to the Island of Oahu. When she was forced by Namakaokahai to flee again, Pele moved southeast to Maui and finally to Hawaii, where she now lives in the Halemaumau Crater at the summit of Kilauea Volcano. The mythical flight of Pele from Kauai to Hawaii, which alludes to the eternal struggle between the growth of volcanic islands from eruptions and their later erosion by ocean waves, is consistent with geologic evidence obtained centuries later that clearly shows the islands becoming younger from northwest to southeast.

FROM LEGEND TO FACT

The oldest of the Hawaiian Islands is a submerged group known as the Emperor Seamount chain that runs southward in a straight line from the eastern tip of the Aleutian Islands. Their basaltic rock is about 65 million years old. Extending from these islands is another line of mostly submerged volcanoes and flat-topped mountains called guyots; together they are known as the Hawaiian chain. The Midway Islands are at the northwestern end of the chain. Their volcanic rock is much younger, only about 27 million years old.

Southeast of the Midway chain, the next visible island is Kauai, the oldest of the large islands that make up the Hawaiian Islands. Its lavas are between four and six million years old. The next youngest in line is the

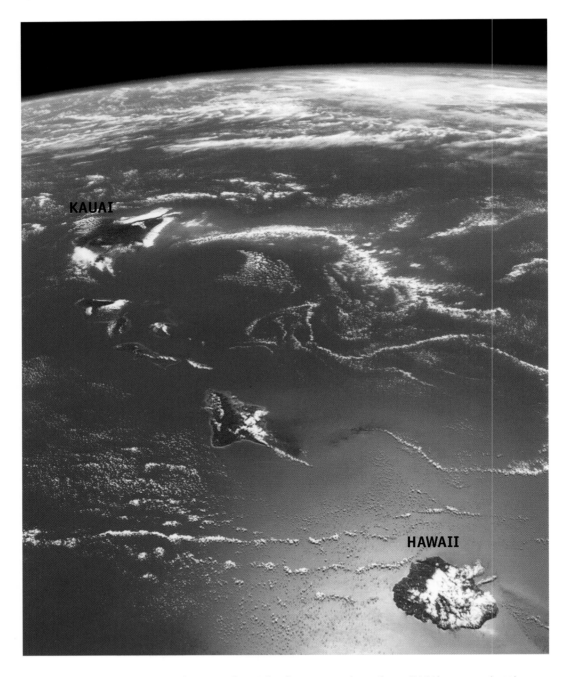

The volcanic chain forming the Hawaiian Islands, as seen here from NASA's space shuttle, occurs not at the plate edges but near the middle of the Pacific Plate. This was a mystery to scientists until "hotspots" were discovered.

island of Oahu, whose lavas are between two and three million years old. It is followed by Molokai, younger still, at only one to two million years. Then comes Maui, whose lavas are even younger, less than a million years old. Today the lava flows of Hawaii, the largest island of the chain, are fed by the active volcanoes of Mauna Loa and Kilauea, which pour forth enormous amounts of new lava. Those two volcanoes will probably remain active for another half million years or so before they, too, become dormant relics.

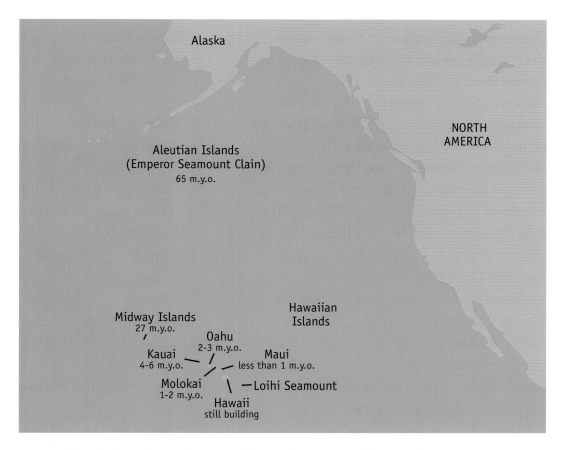

Hotspot islands formed over the past millions of years popped up one by one beginning in the region of the Aleutian Islands (oldest) then southward to the Midway Islands, then southeast to the Hawaiian chain. The island chain formed, and continues to form, as a hotspot in Earth's mantle pushes magma up through the crust as the Pacific Plate continues to glide across the hotspot. (m.y.o. = million years old)

But the story of the Hawaiian chain of volcanic islands doesn't end there. A new volcano to the southeast is now boiling up out of the seafloor. Examination of the ocean bottom about 20 miles (32 kilometers) from Hawaii reveals a fresh volcanic heap named the Loihi Seamount. There can be little doubt that still another tropical volcanic island, named Loihi, will someday be added to the Hawaiian chain.

HOTSPOT PLUMES

One reason geologists are fascinated by the story of the Hawaiian Islands is that they provide convincing evidence for a process geologists call the plume theory. They think that hot rock in certain parts of the lower mantle rises as a crown-shaped mass, or plume, into the upper mantle rock and melts its way up through the crust as a broad area called a hotspot. The term was coined by J. Tuzo Wilson, the Canadian geophysicist who devised the theory. He envisioned things this way.

The Hawaiian Islands and Emperor Seamounts chains form a line along the inner region of the Pacific Plate. As the plate continues to move northwest it carries the island chain along with it. For millions of years hot magma has been melting its way up through the moving Pacific Plate as a hotspot. Over time countless eruptions of the hotspot have built up a chain of seamounts that one by one eventually poked above the waves as volcanic islands carried along by the moving Pacific Plate. As the plate carried one island northward, another in the chain formed right behind it. It, too, remained active for two or three million years, and it later died as the Pacific Plate carried it beyond the hotspot. Loihi is the newest link in that long chain of volcanic islands.

Over the past ten million years, there have been as many as one hundred twenty hotspots, according to one study, but only about twenty-five seem to be active now. The African Plate is known to have about half a dozen. Most hotspots are located beneath plate interiors rather than at the edges. Some,

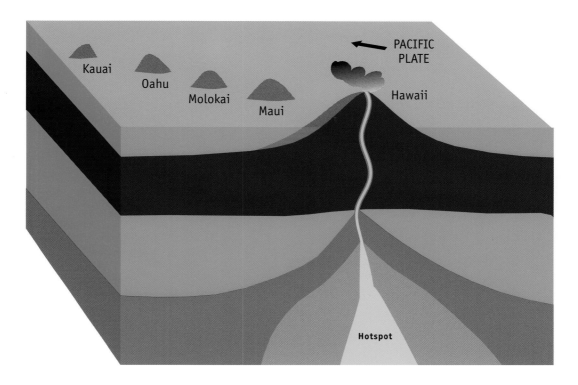

Off and on for millions of years, according to J. Tuzo Wilson's plume theory, a pool of magma deep in Earth's mantle has forced its way up through the Pacific Plate, popping up volcanic mountains as the plate glides generally westward.

however, are found near the mid-ocean ridge system, beneath Iceland and the Azores in the Atlantic, and the Galapagos Islands off the west coast of South America, for example. The North American Plate also seems to have a few hotspots. Most likely there is one beneath the crust in the region of Yellowstone National Park in northwestern Wyoming. The park has several old large craters formed by three gigantic eruptions that occurred over the past two million years. Today heat from Yellowstone's hotspot keeps geysers gushing and lends its warmth to more than 10,000 pools and springs.

Like the Hawaiian Islands, the Galapagos Islands off the western coast of South America were also boiled up out of the seafloor by volcanic eruptions. The islands support a wide variety of animal life forms that fascinated the English naturalist Charles Darwin in the early 1800s. Among them are the flightless cormorant and marine iguanas (above) and giant tortoises (below).

A "hotspot" of upwelling magma beneath the interior of the North American Plate may be the source of Yellowstone National Park's thousands of geysers and hot springs, including geothermal terraces like this one in the Wyoming section of the park.

Geysers galore continuously erupt throughout Kamchatka's remarkable Valley of Geysers. Some two hundred of them hiss, roar, gurgle, and plop as mudpots. The field is the second largest in the world after Yellowstone and is kept active by the grinding together of the Eurasian and Pacific Plates.

When a plume rises into the upper mantle, it fans out to cover an area a few hundred miles across. Each of the twenty-five or so major hotspot plumes around the world creates a zone of volcanic activity. Outside the region of a plume's hot, rising, lightweight rock is an area of cooler, heavyweight mantle rock that is sinking and replacing it. In this way huge circulating cells called convection cells seem to be formed. Whether or not the theory of convection cells is correct, plumes and hotspots exist and every day the force of their heat is changing the face of our not-so-solid Earth.

Above: Many volcanoes have a circular, water-filled basin called a caldera. Some calderas, in spite of their great beauty, are lifeless pools of strongly acidic water. Drop a nail into this one in Kamchatka, for example, and two days later the nail has been dissolved.

Left: Is Yellowstone's famous geyser Old Faithful becoming Old Unfaithful? It seems to be, since the time between successive discharges of water may be gradually growing longer. Nature's plumbing system, which feeds Old Faithful and other geysers, eventually changes and so alters the character of any geyser.

SIX

CONTINENTAL UPS AND DOWNS

There is more to plate tectonics than just the movement of plates as they slide along like colliding ice floes. Some huge areas of land actually bob up and down, such as Australia, and others are slowly being elevated, such as southern Africa, which has risen 1,000 feet (3,000 meters) over the past 20 million years. Still other regions, such as Indonesia, are sinking. Today, only Indonesia's highest peaks poke up out of the Pacific Ocean. Even our continent, North America, has sunk thousands of feet at least once and then popped back up again. As with the horizontal motion of the planet's plates, the forces that drive the up-and-down movements of continent-size chunks of land lie deep inside Earth. Scientists are just beginning to learn about them.

UP WITH AFRICA

Southern Africa consists of one of the world's most extensive plateaus. It stretches for more than 1,000 miles (1,600 kilometers) and rises to a height of nearly 1 mile (1.6 kilometers). Along with the surrounding ocean floor, southern Africa has been gradually growing higher over the past 100 million years, and it is still rising as a formation geologists call a *superswell*. Why? The region hasn't had a collision of plates for almost 400 million years.

The answer most likely lies deeper than plate boundaries and in forces different from those that push plates sideways. Investigators felt that the place to look was in the upper mantle and asthenosphere, from a depth of about 125 miles (200 kilometers) all the way to the outer core, about 1,800 miles (2,900 kilometers) down. Similar to the mantle rock, the asthenosphere behaves something like a white-hot poker, which can be pressed, bent, and reshaped.

Southern Africa is on the rise as Indonesia slowly sinks. Scientists now think that a massive region of hot rock, called a superswell, deep within the planet is gradually expanding and lifting southern Africa skyward. In the last 20 million years the region has risen some 1,000 feet (305 meters). Indonesia's sinking seems to be caused by cool plate material below growing denser as it cools and therefore sinks.

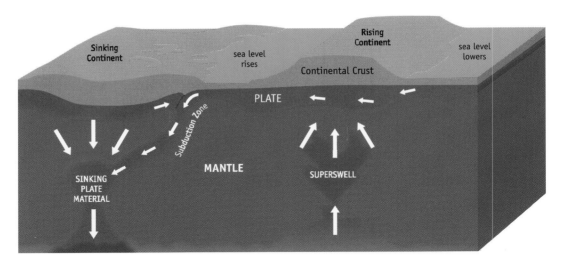

THE UPS AND DOWNS OF CONTINENTS

At left, relatively cool and sinking plate material fed by a subduction zone does not mix with surrounding relatively hot material. As the cool and denser subducted material sinks, it pulls down the continent above and so causes a rise in sea level.

At right, hot material from a superswell originating near the planet's outer core rises as it expands its way up through the cooler and less dense surrounding material. The rising superswell pushes on and lifts the continent above, causing sea level to lower.

The asthenosphere and mantle rock respond to heat and pressure by slowly flowing from one place to another, sometimes being driven by deep heat and churning like a pot of simmering fudge. Hot magma rises because, on being heated by the planet's superhot core of molten iron, it becomes less dense and lighter than the surrounding magma. Then on rising, the lightweight magma cools, becomes denser and heavier, and sinks. Back at the bottom, it is once again heated by the core and rises once more higher into the upper mantle. At least, this is the process scientists have imagined.

Detecting the up and down motions of large masses of magma has been based on a study of the travel times of earthquake waves generated from a location in the crust, then pulsing through the mantle rock, and finally being detected at an earthquake station thousands of miles away. Because seismic waves travel more slowly through hot and less dense rock than through cooler and more dense rock, scientists were able to identify gigantic pockets of especially hot magma and learn about both their density and temperature. In that way it became possible to draw a heat and density map of the mantle rock.

The largest single pocket of superhot magma yet discovered turned out to be under southern Africa. It was shaped like an enormous mushroom cap several thousand miles across and rose about 900 miles (1,450 kilometers) upward from the core. According to American geophysicist Michael Gurnis of the California Institute of Technology, who has studied the matter, "[in the year 2000], we found that the blob is indeed buoyant enough to rise slowly within the mantle—and strong enough to push Africa upward as it goes."

DOWN WITH AUSTRALIA

At first, the discovery of Australia's rise and fall in the early 1970s baffled most geologists. It was not bobbing rapidly but very slowly over millions of years. At present the continent is sinking, right along with Indonesia. About 130 million years ago the eastern half of Australia was under a shallow sea, but by about 70 million years ago the continent popped back up again. Since then it has sunk by about 600 feet (180 meters). Geophysicists couldn't come up with a satisfactory explanation by looking to plate activity for an answer. The movement of plates doesn't seem to lift or lower the land beyond a distance of about 100 miles (160 kilometers) from a plate's edges. Some other force had to be causing these more extensive up and down shifts.

A clue to the puzzle came in the late 1980s, when geologists were studying the high region around Denver, Colorado, which is hundreds of miles away from the edges of colliding plates. Although Denver is about 1 mile (1.6 kilometers) above sea level today, it sits on top of old seafloor sediments laid down some 100 million years ago, when the region was flooded by a shallow sea.

A PEEK INTO THE FUTURE?

What may Earth be like 50 million years from now? Based on what they know about plate movement, geologists can make some educated guesses about how the planet will change.

Antarctica will most likely stay close to where it is today, though it may twist a bit clockwise. Both the Atlantic and Indian Oceans will broaden due to seafloor spreading. At the same time the Pacific Ocean will shrink. Australia may be dragged even deeper by Indonesia's slow sinking. A large chunk of eastern Africa will split off along the Great Rift Valley, sliding northward and creating the world's next major ocean. Meanwhile Africa's general drift northward will close up the Mediterranean Sea into a mere lake, or possibly give rise to a chain of Mediterranean Mountains. In the United States, Baja California and part of California itself will break away along the San Andreas Fault and drift westward out to sea. In only 10 million years from now Los Angeles will be up beside San Francisco, although it will still be attached to the mainland. Some 60 million years after that, Los Angeles will begin to slide down into the Aleutian Trench.

For the truly long-range future, the picture is even more interesting. Over the next few billion years Earth will lose so much of the heat now flowing out of the core and up into the mantle that the mantle's great convection cells may slow to a halt. A cooling and quiet mantle will end plate movement, which in turn will tend to do away with earthquakes and extinguish all but a few volcanoes. Mountain chains will stop being formed, and erosion will gradually wear away those sharp and majestic peaks that still remain.

What house, this resident of Beng Kuln, Indonesia, seems to be asking after an earthquake in June 2000 in which 1,417 people were killed or injured. Indonesia is commonly shaken by earthquakes because it is a volcanic region. Located along the Java Trench and subduction zone, the nation of numerous islands is gradually sinking into the sea. In 1883 its volcanic island of Krakatau exploded and was destroyed in one of the greatest modern-day eruptions.

They concluded that the land had been sucked down by the sinking of an ancient plate near the west coast of North America. Instead of its edge end dipping down into the mantle, the whole plate seems to have sunk, dragging North America down for it to be flooded. But as the plate continued to descend into the mantle, its gravitational tug on the lighter-weight rock above weakened and so allowed the land to rise again. Traces of the ancient plate were revealed in 1996 by seismic mapping of the mantle beneath North America.

Armed with the Denver evidence, Gurnis and his fellow researchers suspected that Australia's ups and downs might also be caused by an ancient sinking plate. It turned out that about 130 million years ago Australia was slipping eastward over an ancient subduction zone.

It was the sinking of cold plate material beneath Australia into the subduction zone that dragged the continent down and allowed ocean water to rush onto the lowered land. Then, when the subduction ceased, Australia began to drift eastward past the zone. This caused the lightweight continental rock to float higher once again.

Although an ancient subduction zone seemed to explain Australia's sinking and subsequent rising, it did not account for the land's renewed descent today. But that puzzle has also been solved. By about 45 million years ago, Australia had changed its drifting course from an easterly direction to a northerly one, toward Indonesia. Since then the Australian Plate has become the fastest moving of all, speeding along at 3.3 inches (8.44 centimeters) a year. Today Indonesia is a record breaker as well, sinking faster than any other region. The reason is its location at the meeting points of subduction zones in the Pacific and Indian Oceans. "And as Indonesia sinks," says Gurnis, "it pulls Australia down with it."

So beneath Indonesia is cold plate material of dense sinking rock. As it sinks, it pulls the ground above down with it, lowering and flooding the land. This occurs because the sinking section of crust lies directly above the colliding edges of the Indo-Australian and Pacific Plates. As the plates grind against one another, they also trigger earthquakes and volcanoes, which makes Indonesia one of the most active sections on the Pacific Rim of Fire.

Southern Africa is a very different story. Back when the supercontinent of Gondwana existed, southern Africa was nestled right in the center, far from any plate boundaries. The geologically calm conditions beneath southern Africa over the past many millions of years, in the absence of cold and dense sinking plates, have permitted the magma beneath to continually absorb heat welling up from the core region. This gradually led to the formation of the superswell of hot magma that is now rising and pushing up on the rock floor of southern Africa.

Only in the past thirty years have geophysicists come to understand some of the key actions, fueled by changes in heat and pressure, occurring within Earth's mantle. Bit by bit, our understanding of what goes on there is helping to explain the many changes we observe at the planet's surface—undersea upwellings of magma that cause the spreading of the seafloor and plate collisions that thrust up mountain ranges, build island arcs, and energize the Pacific Rim of Fire with thousands of earthquakes and volcanoes. The next thirty years are bound to provide new and improved technology that will give us an even clearer and more detailed view of the fascinating world beneath our feet.

GLOSSARY

asthenosphere A zone within the upper mantle where the rock is pliable and permits movement of the crust; it begins below a depth of about 60 miles (100 kilometers) and extends as deep as 450 miles (700 kilometers).

basalt A relatively dense igneous rock.

continental drift The idea that the present continents once existed as a single supercontinent and that the supercontinent broke into smaller continents, which then "drifted" to their present positions and continue to drift.

convergent boundary A zone where two adjacent tectonic plates push against one another.

divergent boundary A zone where two adjacent tectonic plates pull away from one another.

fault Any deep crack in Earth's crustal rock that permits vertical or horizontal slipping of the two faces.

hotspot A region just beneath Earth's crust where superhot magma melts its way up through the crust and spills out as a volcano.

island arc A rim of volcanic islands formed along a trench by a subducting tectonic plate.

lava Molten rock (magma) when it pours out of a volcano.

lithosphere Earth's rigid outer rock layer that includes the crust and occurs above the asthenosphere.

magma Hot rock under high pressure beneath Earth's crust.

magnetometer An instrument used to detect the magnetic alignment of mineral grains in rock.

mantle The layer of hot rock extending from below Earth's lithosphere down to the core to a depth of about 1,800 miles (2,900 kilometers).

plate tectonics The notion that there are six major "plates" and about a dozen smaller ones that form Earth's crust. The continents, along with sections of the ocean floor, are pushed about like giant rafts of rock due to the movement of the puttylike rock of the asthenosphere.

plume A hot mushroom-cap-shaped region of molten rock that flows from the lower mantle up to the crust and may give rise to a hotspot.

plume theory The idea that hot rock in certain parts of the lower mantle rises as a crown-shaped mass, or plume, into the upper mantle rock and melts its way up through the crust as a broad area called a hotspot.

rift valley A fracture in Earth's crust along which molten rock from the mantle wells up and flows out onto the surrounding land (or seafloor). A feature found in East Africa and the Middle East.

subduction zone A region along which one tectonic plate collides with and descends beneath a neighboring plate into the deep mantle. Part of the depressed plate's edge then melts.

superswell A large region of hot, rising, upper mantle rock the pressure of which is strong enough to push the overlying crust into a plateau a thousand or more feet high.

transform boundary The zone where two plates rub and grind against each other along their common boundary. The zone of contact may run from a few miles to hundreds of miles long, depending on the size of the plates involved.

FURTHER READING

The following books are suitable for young readers who want to learn more about plate tectonics.

Cox, Reg. *The Seven Wonders of the Natural World*. New York: Chelsea House, 2000.

Nicolson, Cynthia Pratt. *Earth Dance: How Volcanoes, Earthquakes, Tidal Waves and Geysers Shake our Restless Planet*. Tonawanda, NY: Kids Can Press, 1999.

Sattler, Helen Roney. *Our Patchwork Planet: The Story of Plate Tectonics*. Chicago: Lothrop Lee & Shepard, 1995.

Watson, Nancy, et al. *Our Violent Earth*. Washington, DC: National Geographic, 1982.

WEBSITES

The following Internet sites offer information about and pictures of plate tectonics, many of them with links to other sites.

http://www.extremescience.com This site, in its "Lesson in Plate Tectonics," discusses how plate tectonics works. It includs a map and many illustrations.

http://www.pbs.org/wgbh/aso/tryit/tectonics This site explains the theory of plate tectonics and provides interactive activity so that the user can also get a "feel" for how continental drift works.

http://pubs.usgs.gov/publications/text/dynamic.html This tells the story of plate tectonics, with illustrations of each of the most important elements.

http://www.scotese.com The site of the PALEOMAP Project. The project's goal is to illustrate the plate tectonic development of the ocean basins and continents, and the changing distribution of land and sea during the last billion years. Includes three-dimensional maps and animations.

http://www.thinkquest.org/library/lib/site_sum_outside.html?tname=17701&url=17701 This site was created by high school students. It includes lots of graphics that discuss and demonstrate the theory of plate tectonics. One feature demonstrates the differences among earthquakes of different magnitudes. The site includes a puzzle of volcanoes, tsunamis, and land formation, as well as a quiz and an interactive game.

http://www.cotf.edu/ete/modules/sese/earthsysflr/plates1.html This site includes maps and text about plate tectonics, and gives details on convergent, divergent, and transform boundaries.

http://www.geotimes.org/current This online earth science news-magazine is published by the American Geological Institute. It includes short and in-depth news articles on issues and events affecting Earth and earth science, from discussions of earthquakes to articles on the death of prominent scientist and science writer, Stephen Jay Gould.

http://www.discovery.com/exp/earthjourneys/earthjourneys.html This website of the American Museum of Natural History is an online exploration of Earth, its history and present state, including photos and text, and even a virtual journey to the center of our planet.

BIBLIOGRAPHY

These are among the resources used in researching this book.

Calder, Nigel. *The Restless Earth*. New York: Viking Press, 1972.

Deitz, Robert S., and John C. Holden. "The Breakup of Pangaea." *Scientific American*, October 1970, pp. 30–41.

Dvorak, John J., Carl Johnson, and Robert I. Tilling. "Dynamics of Kilauea Volcano." *Scientific American*, "Earth from Inside Out," special issue 2000, pp. 16-23.

Gallant, Roy A., *Dance of the Continents*. Tarrytown, NY: Marshall Cavendish, 2000.

Gallant, Roy A., and Christopher J. Schuberth. *Earth: The Making of a Planet*. Tarrytown, NY: Marshall Cavendish, 1998.

Gurnis, Michael. "Sculpting the Earth from Inside Out." *Scientific American*, March 2001, pp. 40–47.

Helffrich, George R., and Bernard J. Wood. "The Earth's Mantle." *Nature*, August 2001, vol. 412, pp. 501–507.

Jephcoat, Andres, and Keith Refson. "Core Beliefs." *Nature*, September 2001, vol. 413, pp. 27–30.

Kearey, Philip, and Frederick J. Vine. *Global Tectonics*. Oxford, England: Blackwell Sciences Ltd., 1996.

Leet, L. Don. *Causes of Catastrophe*. New York: McGraw-Hill, 1948.

Mestel, Rosie. "Mush in the Mantle." *Earth*, February 1997, pp. 22–25.

Milstein, Michael. "Cooking Up a Volcano." *Earth*, April 1998, pp. 24–31.

Montgomery, Carla W. *Physical Geology*. New York: McGraw Hill, 1992.

Pendick, Daniel. "Himalayan High Tension." *Earth*, October 1996, pp. 46–53.

Tarling, Don and Maureen. *Continental Drift*. Garden City, NY: Doubleday, 1971.

Taylor, Ross S., and Scott M. McLennan. "The Evolution of a Continental Crust." *Scientific American* "Revolutions in Science," special issue 1999, pp. 12–17.

United States Geological Survey, *This Dynamic Earth*. http://pubs.usgs.gov.

Vitaliano, Dorothy B. *Legends of the Earth*. Bloomington, IN: Indiana University Press, 1973.

Yogodzinski, G. M., J. M. Lees, T. G. Churikova, F. Dorendorf, G. Wöerner, O. N. Volynets. "Geochemical Evidence for the Melting of Subducting Oceanic Lithosphere at Plate Edges." *Nature*, January 25, 2001, vol. 409, pp. 500–503.

INDEX

Page numbers in **boldface**
are illustrations.

About the Author

Roy A. Gallant, called "one of the deans of American science writers for children" by *School Library Journal*, is the author of almost one hundred books on scientific subjects, including the best-selling National Geographic Society's *Atlas of Our Universe*. Among his many other books are *When the Sun Dies*; *Earth: The Making of a Planet*; *Before the Sun Dies*; *Earth's Vanishing Forests*; *The Day the Sky Split Apart*, which won the 1997 John Burroughs award for nature writing; and *Meteorite Hunter*, a collection of accounts about his expeditions to Siberia to document major meteorite impact crater events.

From 1979 to 2000, (professor emeritus) Gallant was director of the Southworth Planetarium at the University of Southern Maine. He has taught astronomy there and at the Maine College of Art. For several years he was on the staff of New York's American Museum of Natural History and a member of the faculty of the museum's Hayden Planetarium. His specialty is documenting on film and in writing the history of major Siberian meteorite impact sites. To date, he has organized eight expeditions to Russia and is planning his ninth, which will take him into the Altai Mountains near Mongolia. He has written articles about his expeditions for *Sky & Telescope* magazine and for the journal *Meteorite*. Professor Gallant is a fellow of the Royal Astronomical Society of London and a member of the New York Academy of Sciences. He lives in Rangeley, Maine.